FEB 16 93   DATE

APR 13 '92
MAR 15 '93

PEANUT PARADE

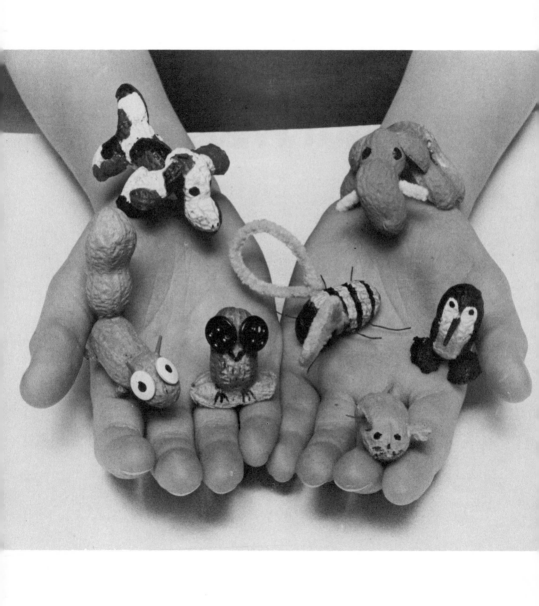

PEANUT PARADE

by Jane Sholinsky

Photographs by Dan S. Nelken

Julian Messner New York

Manufactured in the United States of America

Design by Sheila Lynch

Library of Congress Cataloging in Publication Data

Sholinsky, Jane.
Peanut parade.

SUMMARY: Directions for making animal figures from
peanuts with several suggestions for peanut jewelry.
1. Peanut craft—Juvenile literature.
[1. Peanut craft.   2. Handicraft]   I. Nelken, Dan S.
II. Title.
TT880.S55   745.59′2   78-25852
ISBN 0-671-32944-8

*For Marc's grandma Ceil
who was also Papa's* ginendal

The author would like to extend a
special thank you to her husband, Stephen,
who cheerfully agreed to contribute
the drawings in this book.

Messner Books by Jane Sholinsky

Peanut Parade
In the Saddle:
   Horseback Riding for Girls and Boys

CONTENTS

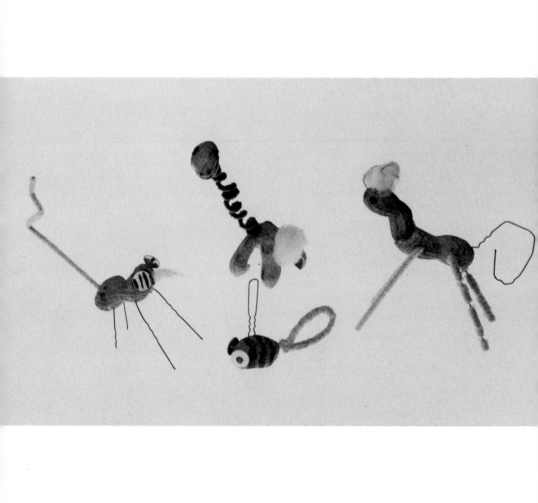

WHAT'S IN A NUT?

Some of nature's most delightful shapes can be found in peanuts. Hold a peanut in your hand. Turn it every possible way. What do you see? Does it make you think of a squirrel, an owl, or perhaps a monkey's head?

Once you really start to look at peanuts, your imagination will take over and you will start seeing all sorts of animal forms. The mere turning of a shell may give you a flash of inspiration, and in no time your creative instincts may turn a peanut into a bumblebee. You needn't be limited to real animals. You might also want to create dinosaurs, monsters, or creatures from another planet.

When creating your peanut parade, don't try to copy nature exactly. Let your own personality, creativity, and artistry show through. Don't be afraid to experiment with the peanuts. The result will be unique and rewarding, yours and yours alone.

Peanut plant

MEET THE PEANUT

The peanut, known as the ''groundnut'' in the rest of the world, is neither a true pea nor a nut. It does, however, grow in the ground. Its closest well known relative is the black-eyed pea, and like it, the peanut produces its seeds in a pod.

As the peanut plant grows, it produces a pretty yellow flower. When the flower wilts, its stem bends over towards the ground and pushes its way into the earth. Once buried, it develops into a pod (a seed container).

The peanut pod usually contains one or two seeds and has a soft shell. It is after the peanut pod is harvested that its shell begins to dry and hardens into what we have come to call the peanut.

GATHERING MATERIALS

The shell of the peanut is easy to use for crafts projects. It is neither too hard nor too soft. It can be easily pierced (punctured with holes using pins or toothpicks), cut, filed, glued, painted, and otherwise decorated.

Listed below are some of the tools and materials you will need for creating your peanut parade. You should be able to find almost everything you need right at home.

Tools

Small scissors—with sharp tips, for cutting shells.

Paint brush.

Round toothpicks—to pierce holes in the peanut shells.

T-pins—for making thin holes. They can also be used to hold a peanut when painting the shell. They are available at the five and ten cent store.

Emery boards—for filing down rough edges, or for parts of animals.

Ruler.

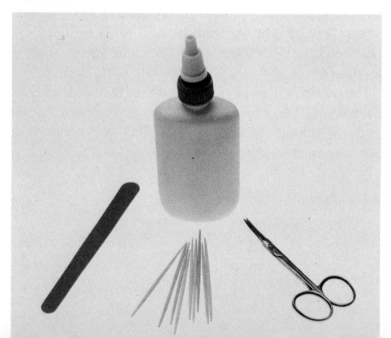

Materials

Peanuts.

Paint—use poster paint or tempera for good colors and
easy cleaning.

Pipe cleaners.

Glue—white plastic type is best. It dries colorless and
will not flake.

Colorless nail polish—for a protective finish for your
peanuts.

Cotton—either on a roll or in balls, for decoration.

Thin wire—in gold or silver. Available at the five and ten
cent store.

Buttons—small ones are best. For eyes.

Hairpins—black, thin. Available at the five and ten cent
store.

Thumbtacks—white are best. Usually used for eyes.

Glitter—comes in small jars, available at the five and ten
cent store.

WORKING WITH PEANUTS

Once you have all your tools and materials, you will want to begin.

Before you start, it is wise to get organized. Here are some helpful hints to keep in mind when crafting your peanut parade.

Peanuts come in an endless variety of shapes and sizes. However, for our purposes, they have been grouped into four general categories of shapes: common, oval, oval-round, and odd.

Take a few minutes to familiarize yourself with the shapes in the photograph. This will make it easier for you to spot the shell shape you will need later.

Separate your peanuts. Pick out the ones with cracked shells and set them aside in a separate bowl. They are used when you need bits of shell. The cracks usually run along the seam of the shell, making it easy to cut the shell into pieces.

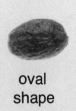

oval
shape

odd
shape

oval-round
shape

common
shape

If a shell should break when you are working on a peanut project, don't discard it. Save the shell for use as a bit piece, and save the nut for a snack.

Have an empty bowl or jar handy when making your peanut parade. Each time you split a shell or break one, remove the kernel and put it in the bowl for eating.

Most shells are easily pierced. However, if you can't start a hole with a toothpick, set that peanut aside and look for another.

If you want to make one or more of the animals shown in this book, be sure to first read the directions carefully. Also experiment with mixing and matching body parts to make creatures all your own.

Now that you are ready to begin, let your imagination take over.

15

Baby Seals

Materials:
1 peanut
emery board
paint—brown, black,
   and white

Tools:
toothpick
scissors
paint brush

These two baby seals, which seem to be resting on ice, are easy to make. All you need is a small, common-shape shell for each body and pieces of emery board for the wings.

To make a baby seal:

1. Cut a 1/2-inch piece off each end of the emery board.

2. Using a toothpick, pierce a hole in the middle of the shell on each side. Insert the tip of your scissors into each hole and cut a slit along the length of the shell.

3. Now insert one piece of 1/8-inch emery board into each side of the shell and tilt it downward.

4. Paint your seal brown. Once the paint dries, paint two large, round, black eyes and circle them in white.

Snake

Materials:
5 peanuts
pipe cleaner
paint—colors you select

Tools:
toothpicks
paint brush

All snakes have the same kind of shape—long, round, and legless. They zigzag over the ground, leaving a wiggly trail. You can make a snake that does the same thing with five common-shape peanuts.

1. Try to find five shells that are about the same size. To make the snake's head, pierce a hole with a toothpick in the end of one shell. Then insert a pipe cleaner 1/4-inch into the hole.

2. Now take another peanut and push a toothpick through the entire length of the shell so that the tip comes out the other end. Remove the toothpick. Then string the peanut on the pipe cleaner.

3. Do the same with two more peanut shells. Then leave 1/4-inch of the pipe cleaner sticking out from the fourth peanut and cut off the rest.

4. The fifth shell will be the tail of your snake. With a toothpick, pierce a hole in only one end of the last peanut. Push the pipe cleaner 1/4-inch into the hole. Now you can bend the snake into different positions and move it along the table or floor.

What kind of snake will you paint? A garter snake with lengthwise stripes? A black racer snake or a coral snake like the one shown here? Don't forget to add the eyes.

Spider

Materials:
1 peanut
4 hairpins—black
paint—black

Tools:
toothpick
ruler
paint brush

One of the ways to turn a peanut into a spider is to add eight legs.

    1. Look for an oval-shape peanut for the spider's body. Then, with a toothpick, make four holes right through the body of the shell near the bottom. Try to space the four sets of holes evenly.

    2. Now open four hairpins and insert one through each of the four holes. These are the spider's legs. Be sure the legs are even on both sides of the shell, then bend them into position.

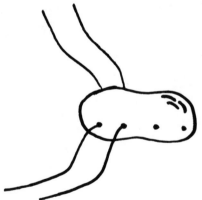

3. Paint your spider the same color as the hairpins. Then send it off to spin a web.

You can draw a web by tracing the one in the picture. Or, using a ruler, draw eight lines that cross one another at a central point. Then connect these lines with curved ones like you see in the photograph.

Mouse

Materials:
1 peanut
bits of shell
toothpick
cotton
glue
paint—grey and black

Tools:
toothpick
emery board
paint brush

A field mouse can be found in the most unlikely places. To make one, you need a large, oval-shape peanut. Choose one from your collection. Turn the shell around in your hand to decide which end is wider and best for the mouse's head.

1. Bits of shell are all you need for the ears. Choose two pieces from your collection and cut each into a small, round shape. With your emery board, file one side of each ear until it is flat. Then glue the flat sides to each side of the mouse's head. Let the glue dry thoroughly.

2. For the back, pierce a hole with a toothpick in the other end of the shell. Then cut the toothpick in half and insert one of the halves into the shell for the mouse's tail. Be sure to position the toothpick with its smooth, pointed end facing out.

3. Paint your mouse grey. After that coat dries, paint two black eyes and a black nose. To give your mouse whiskers, glue a thin strip of cotton just below the nose.

Penguins

Materials:
1 peanut
bits of shell
toothpick
paint—white and black
colorless nail polish
glue

Tools:
toothpick
emery board
paint brush

Seen from far away, these three penguins might easily be taken for a trio of little men. Like many people, these birds stand erect and are flat-footed. They often walk in straight lines like soldiers.

To make a penguin, look for a common-shape peanut that shows a small, natural, beak-like shape at one end. This will be your penguin's head.

1. With a toothpick, pierce a hole in the shell at the tip of the beak-like shape. Then cut off 1/3 of the toothpick and insert it into the hole with the pointed end facing out. If necessary, glue the toothpick into position.

2. To make feet for your penguin, you will need two large pieces from your collection of shell bits. Cut the two pieces into round shapes, and, with your emery board, file the edges smooth so that each foot will stand firm and not wobble.

3. When gluing the feet into position, take time to stand the penguin so that it is well-balanced. Also, as you glue each foot to the bottom of the penguin's body, hold the shells together for a few minutes. Then stand the penguin upright, and let the glue dry thoroughly.

4. After you paint your penguin white and black like the ones shown here, add two small black dots for eyes. When the paint is dry, cover the penguin with a coat of colorless nail polish to make it appear to be wet.

Cat

Materials:
2 peanuts
toothpick
thin wire
paint—color you
    select, or colorless
    nail polish
glitter—green
glue

Tools:
toothpick
T-pin
scissors
emery board
paint brush

The cat is probably one of the most beautiful and graceful of animals. As it moves, the muscles of its body ripple smoothly under its soft fur. Even when at rest, every line of its body curves gracefully.

A common-shape peanut is used to make the body of this cat. If you want to make a cat standing or stalking a bird, just add legs to the body shell and change the position of the tail.

1. When looking for a common-shape peanut for the cat's head and body, see if you can find one that suggests the roundness of a cat's nose. Then use a toothpick to make two holes on the top of the shell for ears. Cut 1/4-inch off both ends of the toothpick, and insert both cut-off ends into the two holes, pointed ends up. Glue into place.

2. Your cat also needs whiskers. To make these, cut six 1/2-inch pieces of thin wire. Then use a T-pin to make three tiny holes on either side of the cat's nose. Insert one piece of wire into each hole.

3. For your cat's tail, you will need a large, common-shape shell. See if you can find one in your collection of shell bits. If not, look for a large, common-shape peanut with a cracked shell. Then insert the tip of your scissors into the crack and snip along the length cutting the shell into two halves. Set aside the unused half for another project.

4. Now cut one end of the half-shell on a slant and file it smooth with your emery board so that it fits against the back of the body shell. Then glue the tail and body shells together. Let the glue dry thoroughly.

5. You can paint your cat the color of your choice. If you want, you can brush on colorless nail polish instead of painting your cat. Whichever you choose, glue on green glitter for eyes. Then place your cat on a cotton pillow.

Turtle

Materials:
2 peanuts
pipe cleaner
paint—green and black
glue

Tools:
scissors
emery board
toothpicks
paint brush

1. To begin crafting your turtle, look through your collection of peanuts for one with a cracked shell. Any shape shell will do. Then insert the tip of your scissors into the crack and snip along the length, cutting the shell into two halves.

2. One at a time, cut each half-shell into six small, oval-shape pieces. Round off the edges with your emery board.

3. Now look for an oval-shape peanut for your turtle's body. Then choose one of the cut pieces of shell for the turtle's head, and glue it to the bottom of the body shell at the front. Let the glue dry thoroughly.

4. The remaining pieces of shell will be used to make the turtle's back. Starting at the sides of the body, glue the pieces to the turtle's back in a circle. Keep working in circles until you reach the top. Let the glue dry thoroughly.

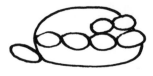

5. You will need a pipe cleaner to make the turtle's feet and tail. With a toothpick, make two sets of holes at the bottom of the body, one set near the front and the other set near the back. Then cut two one-inch pieces of pipe cleaner for the turtle's legs. Insert one of the pieces through the front holes and one through the back holes.

6. Use a toothpick to pierce a hole in the back of the body for the tail. Then cut a 1/2-inch piece of pipe cleaner and insert it halfway into the hole.

7. Paint your turtle green, and add two little black eyes.
Look at your turtle carefully. What kind of turtle does it look like? A box turtle has a high, round shell that it can close up like a box. A map turtle has a wide, flat shell with bumpy edges. A soft-shelled turtle is round and flat and looks like a green pancake.

Butterfly

Materials:
1 peanut
pipe cleaner
cotton
thin wire
paint—dark brown and
   others you select
glue

Tools:
T-pin
scissors
paint brush

1. The butterfly is a delicate and graceful creature. To make one, look for a common-shape peanut that can stand without toppling to its side. Then paint the shell dark brown and set it aside to dry.

2. The eyes on the butterfly shown here are two small pieces of cotton. Roll each piece in the palm of your hand to make tiny round eyes. Then glue each eye to the shell.

3. To make the butterfly's feelers, use a T-pin to pierce two holes in the shell at the top of the area between the eyes. Then cut two one-inch pieces of thin wire and insert one piece into each hole.

4. Use a pipe cleaner to make a frame for your butterfly's wings. Bend both ends of the pipe cleaner towards the center, making two large loops. Then twist each loop closed at the center. Next, stretch cotton across each loop, and glue it to the pipe cleaner frame. Trim off any extra.

To attach the wings to the butterfly, bend the wings into shape. Then glue them to the back of your peanut shell at the center.

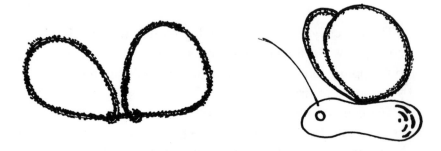

5. The most easily recognized butterfly is the Monarch with its bright orange and black wing pattern. Most butterflies have brightly colored wings. If you are familiar with the various types of butterflies and their wing patterns, you can copy one. If not, use your imagination to paint a bright and colorful design on the butterfly's wings.

Owl on a Perch

Materials:
2 peanuts
2 buttons
paint—grey and brown
glue

Tools:
toothpick
scissors
emery board
paint brush

This owl seems to be sitting on its perch waiting for night to come. That's when the owl hunts for food. Most owls stay up all night and sleep during the day. Would you like to do that?

1. To make an owl, first look for a long, smooth shell for the owl's perch.

Now use a toothpick to pierce a hole in one end of the shell. Then insert the tip of your scissors into the hole and snip along the length, cutting the shell into two halves. Use your emery board to file the bottom edges of the half-shell until it stands without wobbling.

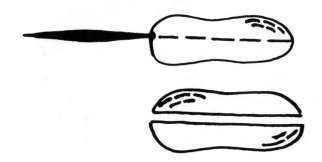

To finish the perch, pierce a hole with the tip of your scissor through the center of the half-shell, and cut out a small piece.

2. For the owl, choose a common-shape peanut and stand it in the hole you've just made. If it doesn't fit, make the hole larger.

Spread glue around the rim of the hole and set the owl into it. Let the glue dry thoroughly.

3. Make the owl's eyes with two buttons. Glue one button to either side of the shell near the top.

4. Paint the owl's body grey and brown. Follow the natural texture of the peanut shell when painting the owl. This helps you get the V-shape design of an owl's feathers.

Giraffe

Materials:
3 peanuts
pipe cleaner—yellow
toothpicks
paint—yellow and brown
glue

Tools:
scissors
toothpicks
paint brush

Here is the tallest of all animals. Its very long neck enables the giraffe to reach high into a tree for leaves. In fact, its neck and legs are so long a giraffe cannot bend down to eat leaves off the ground!

1. To make a giraffe, choose a common-shape peanut for the body shell. With a toothpick, make two holes right through the shell, one for the front legs and one for the hind legs. Insert one pipe cleaner through the front holes and one through the back holes. Then bend the ends to make feet so that your giraffe can stand.

2. Half of a small, common-shape peanut can be used to make the giraffe's tail. Perhaps you already have one in your collection of shell bits. If not, look for a whole peanut with a cracked shell. Insert the tip of your scissors into the crack and snip along the length, cutting it into two halves. Then glue the half-shell to the back of the giraffe's body. Let the glue dry thoroughly.

3. Use a pipe cleaner for the giraffe's long neck. First make a hole in the top of the body at the front with a toothpick. Then cut a pipe cleaner in half, and insert one end into the hole.

4. Now look for a very small, common-shape peanut to be the giraffe's head. With a toothpick, make two holes on the top for ears. Then cut 1/4-inch off both ends of the toothpick and insert one into each hole. Position the tooth-picks with their pointed ends going into the shell and glue into place.

5. To attach the head to the neck, make a hole with a toothpick in the bottom of the shell under the ears. Insert the end of the long pipe cleaner neck.

6. You can paint your giraffe yellow. Once the paint is dry, add brownish colored spots and squiggles to the giraffe's body and head. The spots, which look like leaves, help a giraffe hide from its enemies. When a giraffe goes into an area with trees, its enemies can't tell if they're looking at leaves or at the giraffe.

Bumblebee

Materials:
1 peanut
pipe cleaner—yellow
thin wire
paint—black and yellow

Tools:
toothpick
scissors
T-pin
paint brush

This bumblebee seems to be searching for nectar. Bees also carry pollen or seeds from flower to flower and so help flowers to reproduce. The bee is a helpful insect, although not a friendly one. However, you can make a bumblebee that won't sting.

1. Choose a large, round peanut with a hump-like shape. Paint the bee's head black and the body yellow. Once that paint dries, paint black stripes on the bee's body.

2. With a T-pin, make holes for legs through the bottom sides of the peanut shell: one pair near the front, one pair near the center, and the third pair near the back. Then cut three one-inch pieces of thin wire, and insert one piece of wire through each of the pairs of holes. Be sure the legs are even on both sides of the shell.

3. Wings for your bumblebee are made with a yellow pipe cleaner. Cut the pipe cleaner in half. Then bend each half into a loop, leaving a little extra pipe cleaner at one end. With the other end, twist the loop closed.

To attach the wings to the body, use a toothpick to make a hole on each side of the shell near the top. Then insert the tip of each pipe cleaner wing, and send your bumblebee off in search of nectar in the heart of a paper flower.

Rabbit

Materials:
2 peanuts
pipe cleaner
cotton
toothpicks
thin wire
paint—grey, white, brown,
  or black
glue

Tools:
toothpicks
T-pin
paint brush

A rabbit is a soft and furry animal. Your peanut rabbit may not be so soft, but it's easy to make.

1. Look in your collection for an oval-round shape peanut shell for the rabbit's head. With a toothpick, make two holes on the top of the shell for ears. Then cut the toothpick in half and insert each half in the holes. Be sure the toothpicks have their pointed ends facing up before you glue them into position.

2. Now choose a common-shape peanut shell for the body. With a toothpick, make two holes right through the shell, one for the front legs and one for the hind legs. Cut a one-inch piece of pipe cleaner and insert it through the front holes. Cut a three-inch piece of pipe cleaner and insert it through the back holes for hind legs (a rabbit's hind legs are longer than its front legs). Bend the hind legs into shape.

3. To attach the head and body shells, glue the head shell to the front of the body. Hold the shells together for a few minutes until they stick. Let the glue dry thoroughly.

4. The rabbit shown here is painted grey. However, you can paint your rabbit white, brown, or black.

5. Once the paint dries, paint black eyes and a black nose. Then roll a piece of cotton into a ball for the rabbit's tail and glue it to the back of the body shell.

6. Your rabbit also needs whiskers. Use a T-pin to pierce three tiny holes on either side of the nose. Then cut six 3/4-inch pieces of thin wire and insert one into each hole.

Rooster

Materials: **Tools:**
1 peanut   toothpicks
2 buttons   scissors
toothpicks   paint brush
thin wire
cotton
hairpin
glue
paint—color you select

You can easily make a rooster by using a large, common-shape peanut with a slight curve at the top.

  1. Look in your collection for such a peanut. Turn the shell this way and that until you can see the shape of the rooster's proud head. With a toothpick, make a hole where the rooster's beak should be. Then cut off 1/4-inch of the toothpick and insert the piece into the hole. Be sure the pointed end is facing out before you glue it into position.

2. For your rooster's tail, you will need a dozen or so pieces of thin wire. Cut each piece of wire the same three-inch length. Hold all the wires together at one end and then twist those ends together.

3. With a toothpick, make a hole for the tail in the back of the shell near the bottom. Then insert the twisted ends of the wire into the hole and glue into position. Once the glue has dried thoroughly, fan the tail out as shown on this roost-er.

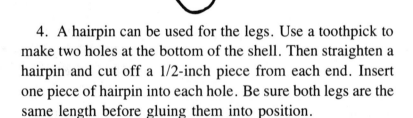

4. A hairpin can be used for the legs. Use a toothpick to make two holes at the bottom of the shell. Then straighten a hairpin and cut off a 1/2-inch piece from each end. Insert one piece of hairpin into each hole. Be sure both legs are the same length before gluing them into position.

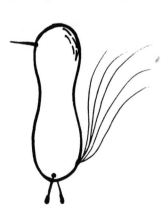

5. To make the comb on your rooster's head, glue a small piece of cotton to the top of the shell.

6. Attach two buttons for eyes just below either side of the comb.

7. Paint your rooster. Then set it down to crow in a new day.

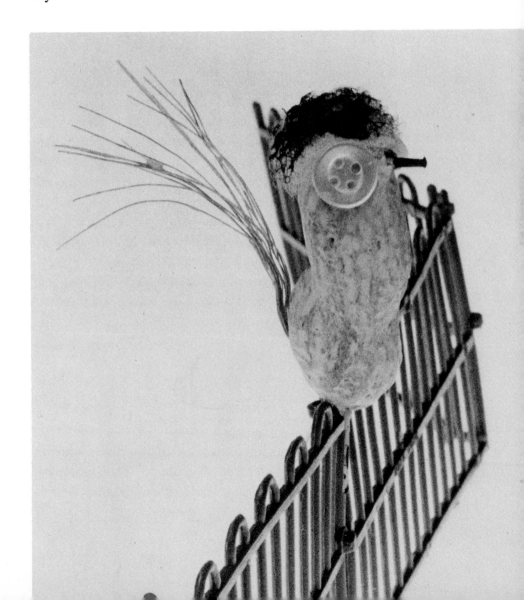

Housefly

Materials:
1 peanut
4 hairpins
2 thumbtacks
paint—grey and red
glue

Tools:
toothpick
scissors
paint brush

The housefly can be a real pest, especially when it lands on your food or tickles you as it walks on your skin. Even so, flies are fun to make from peanuts.

1. Choose a large, oval-shape peanut from your collection. With a toothpick, make three holes at the bottom through both sides of the shell, one near the front, one near the center, and one near the back.

2. Cut two hairpins in half with your scissors. If you don't have sharp scissors, try bending the hairpins back and forth from the center until they break.

Insert one piece of hairpin through each of the three holes for legs. Make sure the legs are even before bending them into shape.

3. To make wings for your housefly, take two more hairpins and bend them into an oval shape. Then make two holes with a toothpick, one on each side of the shell near the

top. Insert both ends of each hairpin into each hole, and spread glue around the hole to keep the wings in position.

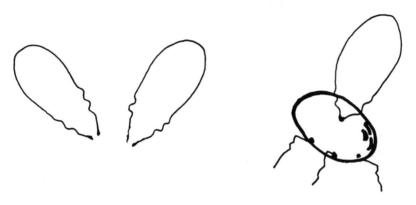

4. Once the glue has dried thoroughly, paint your house-fly grey.

5. For eyes, push a thumbtack into either side of the front of the shell. Then paint each thumbtack red.

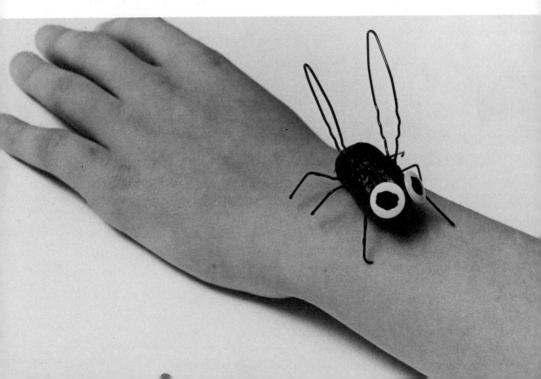

Monkey

Materials:
3 peanuts
2 small buttons
2 pipe cleaners
glue

Tools:
scissors
emery board
toothpicks

These playful little monkeys seem to be enjoying them-selves, using their long tails to swing through the trees.

1. To make a monkey, first look for an oval-round shape peanut for the head. Choose one that suggests the shape of a monkey's prominent jaw.

2. Put the head aside while you choose a peanut with a cracked shell for the monkey's ears.

Insert the tip of your scissors into the crack and snip along the length of the shell until you've cut two halves. Then cut two small, round shapes for ears from the other half-shell. With your emery board, file one side of each ear until it is flat. One at a time, glue the flat side of each ear to each side of the monkey's head at eye level. Hold the shells together for a few minutes until they stick. Then let the glue dry thoroughly.

3. The eyes on the monkeys shown here are made from buttons. Choose two small buttons. Then glue them to each side of the head just below the ears.

4. Now select a common-shape peanut for the body. Using a toothpick, pierce holes for arms and legs.

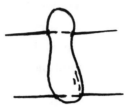

To make the monkey's legs, insert a pipe cleaner through the bottom holes. Then make sure the legs are even on both sides of the shell.

The monkey's arms are shorter than its legs. Cut a pipe cleaner in half, and insert it through the top holes in the shell. Again, make sure the arms are both the same length.

5. Use the other half of the pipe cleaner for the monkey's tail. With a toothpick, make a hole in the back of the body. Insert the pipe cleaner 1/4-inch into the hole and glue it into position.

6. Attach the head and body shells with glue. Once the glue has dried thoroughly, bend the ends of the pipe cleaners to make hands and feet. Then curl the end of the monkey's tail and let it swing.

Dog

Materials:

4 peanuts
bits of shell
paint—black, brown,
   and white, or
   colors you select
glue

Tools:

scissors
toothpick
emery board
paint brush

Decide on the type of dog you want to make. Will it be a Poodle? A Beagle? A Hound Dog? Then look for a peanut that suggests the dog's head. A Terrier has a small face with a square nose. The Beagle's head is wider and its nose longer than a Terrier's. The Hound Dog's head is large with a full jowl.

The dog here is a Beagle. To craft another breed, all you have to do is change the size and shape of the ears and tail.

1. Once you decide on the head shell, look for a common-shape shell for the dog's body. Then use your scissors to cut a small, round hole at the top end of the body where the head will be connected.

2. Glue the head and body shells together. Hold the shells together for a few minutes until they stick. Then let the glue dry thoroughly.

3. Now choose a small, common-shape peanut for your Beagle's ears. With a toothpick, pierce a hole in one end of the shell. Then insert the tip of your scissors into the hole and snip along the length, cutting the shell into two halves.

Cut off one end of each half-shell, and with your emery board, file it smooth so that it fits against the head shell. Then glue one ear to each side of the dog's head. Let the glue dry thoroughly.

4. Small shells with only one nut in them were used to make the legs for this Beagle. Larger shells can be used if you want to make longer legs.

Cut each of two shells into halves the same way you did for the ears. Then cut off one end of each shell and file it smooth.

Glue one leg to each side of the body. Hold the shells together for a few minutes, and then set your dog aside to let the glue dry thoroughly.

5. Your dog's tail can be made from bits of shell. See if you can find a piece of shell in the shape you want. Then use your emery board to file one end so that it fits against the back end of the body. Glue the tail and body shells ·together and let the glue dry thoroughly.

6. The Beagle shown here is painted black, brown, and white. You can paint your dog any color. Once it is dry, paint two large, black eyes to complete your new pet.

Bird in a Nest

Materials:
2 peanuts
bits of shell
hairpins
toothpicks
glue
paint—colors you select

Tools:
scissors
emery board
toothpicks
paint brush

Do you know what makes a bird different from any other animal? Its feathers! All birds have feathers, and birds are the only creatures which do.

1. To make a bird in a nest, first look through your collection for an oval-round shell to make the nest. See if you can find one with a cracked shell. This will make it easier for you to split the shell in half. Insert the tip of your scissors into the crack and snip along the length, cutting the shell into two halves.

With an emery board, file the bottom edges of the half-shell until it stands firmly.

2. With the tip of your scissors, pierce a hole through the center of the half-shell and cut out a small circle.

3. The bird shown here was made from a common-shape peanut. See if you can find one that shows a small, natural, beak-like shape at one end. Then stand it in the hole you have cut in the nest. If it doesn't fit, make the hole larger.

Spread glue around the rim of the hole and set the bird in place. Then let the glue dry thoroughly.

4. You can often tell a bird by the shape of its beak. This bird's toothpick-beak is similar to the long beak of the hummingbird. To make a beak, first use a toothpick to make a hole in the bird's shell where the beak belongs. Then cut off 1/3 of the toothpick, and insert it in the hole with the pointed end facing out. If necessary, glue it into position.

5. Wings for your bird can be made with hairpins. Take two hairpins and bend them into diamond shapes. With a toothpick, make two holes, one in each side of the shell just below the bird's head. Hold the ends of the hairpins together and insert them into the holes. To keep the wings in position, put some glue around each hole. Let the glue dry thoroughly.

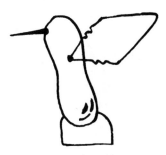

6. To finish the bird's nest, cut the remaining half-shell into six small pieces and round off any rough edges with the emery board. Then glue each piece of shell onto the half-shell. Once the glue has dried, paint your bird in a nest any colors you select.

Squirrel

Materials:
2 peanuts
toothpicks
pipe cleaner
2 thumbtacks
glue
paint—grey and black

Tools:
toothpicks
scissors
paint brush

Squirrels are the little animals which love nuts so much. Every fall, they store away nuts for winter food.

1. To make a squirrel, look for an odd-shape peanut that suggests a squirrel's head and body. Be sure that the part of the shell you will use as the head has enough space for a squirrel's eyes and ears.

2. With a toothpick, pierce a hole for the ears on each side of the shell at the top. Then cut off 1/4-inch from both ends of the toothpick, and insert one end into each hole. Be sure the toothpicks have their pointed ends facing up before you glue them into position.

3. Now use a toothpick to make two holes right through the body part of the shell, one set for the front legs and one set for the hind legs. Cut two one-inch pieces of pipe cleaner and insert one piece through the front holes and one piece through the rear holes. Bend the legs into shape.

63

4. You will need a large, common-shape peanut for your squirrel's tail. However, before you glue it to the body, you need to remove the peanuts from inside. Without the nuts, the shell will be lighter and the squirrel won't topple over when the tail is attached. Use a toothpick to pierce a hole in one end of the tail shell. Then insert the tip of your scissors into the hole and snip around the seam of the shell, cutting it into two halves. Remove the peanuts and place them in your snack bowl for later. Then glue the two halves together.

To attach the tail to the body, glue one end of the tail shell to the back of the squirrel's body. Hold the shells together for a few minutes. Then let the glue dry thoroughly.

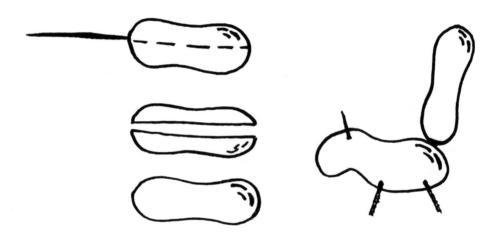

5. Once the glue has dried, paint your squirrel grey. For the eyes, push a thumbtack into either side of the head and paint one big, black dot on each of them.

Horse

Materials:
2 peanuts
toothpicks
thin wire
glue
paint—colors you select

Tools:
scissors
toothpicks
paint brush

The horse is a proud animal. Its long neck and the way it carries its head give it a quality of beauty as well.

1. To make a horse, you should look for an odd-shape peanut that suggests a horse's head and long neck. Once you find the right shell, use a toothpick to make two holes on the top for ears. Then cut 1/4-inch off both ends of the toothpick, and insert one end into each of the holes. Be sure to position the toothpicks with their pointed ends facing up.

2. The horse's body should be a long, straight peanut shell. Avoid a shell with a bump or a common-shape shell. It will only make the horse look swayback, or sagging.

To attach the head and body shells, first use your scissors to cut off one end of the body shell. This will make it easier for the head to fit against the body shell. Then set the bottom of the neck against the cut section of the body shell and glue them together. Hold the shells together for a few minutes. Let the glue dry thoroughly.

3. To make the legs of the horse, use a toothpick to pierce two holes on the bottom of the body shell near the front for forelegs and two holes near the back for hind legs. Then cut two toothpicks in half and insert the end of each half-toothpick into the four holes. Before you glue them into position, make sure that the toothpicks are all the same length and that their pointed ends are facing out.

4. The horse's tail can be made from thin wire. Cut several one-inch pieces and twist them together at one end.

Then make a hole in the end of the body shell with a toothpick, and insert the twisted ends of the wire into the hole.

5. You are now ready to paint your horse. What breed will it be? Will you paint a golden-yellow Palamino? A Pinto with brown spots? Perhaps your horse will have a large blaze like the one shown here. Maybe it will have a tiny white stocking on one leg.

Elephant

Materials:
5 peanuts
pipe cleaner
paint—grey
colorless nail polish
glue

Tools:
scissors
toothpick
paint brush

Imagine you have a hand on the end of your nose. Would you use it to help you eat? That's what an elephant does. The elephant's trunk is its nose. At the end of the trunk are one or two little bumps that the elephant uses as a type of hand with fingers. An elephant can reach with its trunk to tear leaves from a tree. Then the elephant curls its trunk and stuffs the leaves into its mouth. The elephant can also pick up something as tiny as a peanut with its trunk.

1. To make an elephant, you should first look for an odd-shape shell like the ones used here for the elephant's head and trunk. If you can't find one in your collection,

look for two shells to glue together, one for the head and one for the trunk. Then choose a large, fat shell for your elephant's body.

2. To attach the head and body shells, first use your scissors to cut off one end of the body shell. Then, put glue along the cut edge. Set the end of the head shell against the hole in the body shell. Hold the shells together for a few minutes and then let the glue dry thoroughly.

3. The legs of the elephant can be made using half-shells. See if you have four half-shells the same size. If not, look for two small, common-shape peanuts with cracked shells. Then insert the tip of your scissors into the crack and snip around the seam, cutting the shell into two halves.

4. Now cut one end off each half-shell and, with your emery board, file them smooth so that they fit against the body shell. One at a time, glue each leg to the body shell. Be sure each leg has dried thoroughly before going on to the next one.

5. The size of the shell you choose for your elephant's ears will determine if it's an African or Indian elephant. The African elephant has larger ears than its Indian cousin. In fact, the African elephant is the biggest of all land animals.

An oval-shape shell was used to make the ears on each of the elephants shown here. Cut the shell into two halves the same way you did for the leg shells. Then cut off one end of each half and file it smooth. One at a time, glue the ears to each side of the elephant's head and hold them in place for a few minutes.

6. The elephant's tail and tusks are made with pieces of pipe cleaner. Use a toothpick to pierce a hole in the back of the body shell for the tail. Then cut a 1/2-inch piece of pipe cleaner and insert one end into the hole. For the tusks, make one hole on each side of the elephant's head just above the trunk. Then cut two 1/2-inch pieces of pipe cleaner and insert one piece into each hole.

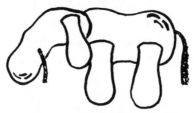

7. Will you paint your elephant? That's up to you. Perhaps you would rather just brush on a coat of colorless nail polish.

PEANUT JEWELRY

As far back as the Stone Age, human beings made jewelry. In their caves, they fashioned necklaces of teeth and animal bones (amulets) to ward off evil spirits. Many people the world over still wear jewelry for luck or because they think it works magic as well as for ornamentation.

Some jewelry is almost priceless. However, much is only costume jewelry, made with imitation gems and inexpensive metals. You can even make costume jewelry from peanuts!

On the following pages are ideas for making peanut jewelry. Use them as stepping-off points to help you get started. Then, once your creative instincts start working, design your own jewelry.

You might surprise a close friend with a beautiful necklace or help cheer up a sick relative with a sparkling pin. Whatever its use, the jewelry will be special because it is yours.

Bracelet

Materials:
4 peanuts
1 pipe cleaner
paint—colors you select
sparkles—colors you select
glue

Tools:
toothpick
paint brush

A bracelet can dress up almost any outfit, especially this one with four colorful, peanut charms.

1. To make a bracelet, look in your collection for four oval-round peanut shells. See if you can find four the same size. If not, look for two small and two large, oval-round peanuts.

2. Decide how you want to arrange the peanuts on your bracelet. Then take one peanut and push a toothpick through the shell until the tip comes out the other end. Do the same with the other three peanuts.

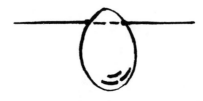

3. Now, one at a time, string the peanuts on the pipe cleaner and space them as shown here.

4. Once your peanuts are arranged on the pipe cleaner, bend the pipe cleaner around your wrist and twist the ends together to make a bracelet. Be sure you leave enough room so that you can slip the bracelet on and off your hand.

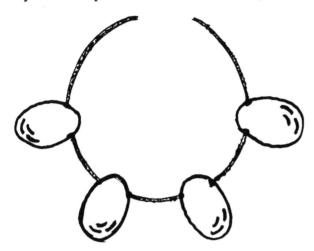

5. Now you are ready to decorate your bracelet. How will it look? The bracelet shown here was painted bright green. After the paint dried, green sparkles were glued to each peanut shell for decoration.

Necklace

Materials:
15 peanuts
3 pipe cleaners
paint—colors you select
glitter—colors you select
glue

Tools:
toothpicks
paint brush

1. Look for fifteen common-shape peanuts about the same size.

2. Take one peanut and use a toothpick to pierce a hole through the length of the shell. Be sure to push the toothpick through the entire length so that the tip comes out the other end. Now insert a pipe cleaner through the holes in the shell.

3. Do the same thing with four other shells. You should be able to string five shells on each of the three pipe cleaners.

4. Once all the shells have been strung, join the pipe cleaners by twisting the ends together to make a circle.

5. Now choose your favorite color or colors to paint your necklace. Once the paint is dry, you can glue on glitter to make your necklace sparkle.

Ring

Materials:
1 peanut
1 pipe cleaner
paint—color you select
glitter—color you select
glue

Tools:
toothpick
scissors
T-pin

You can choose any size or shape peanut shell for your ring form. An unusual shape would make an interesting ring. Perhaps you prefer a small, round shell or a long, narrow form, like the one shown here.

1. Once you have made your choice, use a toothpick to pierce a hole in one end of the shell. Then insert the tip of your scissors into the hole and snip lengthwise, cutting it into two halves.

2. With an emery board, file the edges of the half-shell until they are smooth. Then use a T-pin to pierce a hole right through the sides at the center.

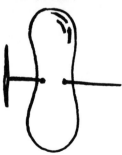

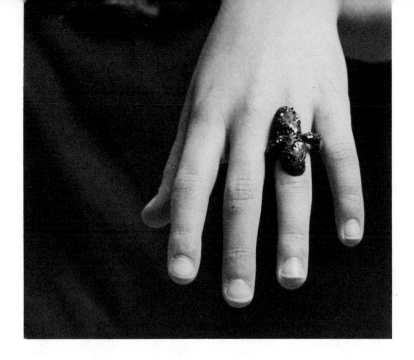

3. Gently push a toothpick through the holes in the shell until the toothpick comes out the other side. This will stretch the size of the hole so you won't have trouble stringing it on a pipe cleaner.

4. Cut a two-inch piece of pipe cleaner and string it through the two holes in the half-shell. Fit the ring to your finger and twist the ends of the pipe cleaner together.

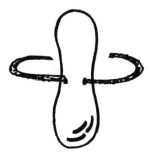

5. Now you can paint the ring the color of your choice and decorate it with glued glitter or beads.

Pin

Materials:
1 peanut
1 pipe cleaner
cardboard
safety pin
glue
paint—color you select
glitter—color you select

Tools:
toothpick
scissors
paint brush

Most of the peanut shapes in your collection can be turned into beautiful pins for you to wear or give as gifts. Select one you feel will make a special design.

1. Use a toothpick to pierce a hole in one end of the shell. Insert the tip of your scissors into the hole and snip along the length, cutting the shell into two halves.

2. To frame your pin, cut a pipe cleaner in half. Then bend one half into the same shape as your half-shell. Trim off any extra. Place the half-shell inside the pipe cleaner frame with the open end facing away from you. Then glue the peanut and frame together.

Once the glue has dried, place the pin on a thin piece of cardboard and outline the shape with a pencil. Remove the shell and cut out the cardboard shape.

To cover the open end of the shell, paste the shape to the back of the pin. Let the glue dry thoroughly.

If you want, you can decorate your pin to match the ring or another piece of jewelry. Or, you can glue glitter around the edge of the pin like the one shown here. Once the paint and decoration is dry, glue a small safety pin to the cardboard backing. Let the glue dry, and you are ready to wear your jewelry.